Name _____   Date _____

HISTORY OF CHEMISTRY

In very early times, people learned how to produce chemical changes. They made these changes work for them long before it was understood why the changes took place. For example, metalsmiths knew how to extract metals from ores. They knew how to combine metals to make jewelry and weapons. People knew how to make alcoholic beverages from fruits and grains. Glassmakers were able to combine and heat substances in order to form them into delicate glass vases and ornaments. All of these skills required the control of chemical changes, but people did not yet understand why the changes occurred.

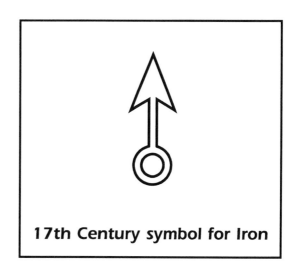

17th Century symbol for Iron

As long as 3000 years ago, people in different countries began developing theories about the nature of the world around them.

They did not, however, have enough knowledge to test their ideas. Early Chinese scholars suggested that all things were made of earth, fire, metal, water, and wood. The Greeks thought everything was made of only four elements: earth, fire, water, and air.

One of the earliest forms of chemistry was alchemy. During the Middle Ages, alchemists spent much of their time trying to turn metals of low value, such as iron and lead, into gold. Although none of them ever achieved that miraculous change, alchemists did make some worthwhile contributions to early chemical knowledge.

Robert Boyle, who lived in the 1600s, is usually considered the first real chemist. He understood that theories should be backed by careful experimentation. In the 1800s, John Dalton of England thought that each element was made of a different kind of atom.

In 1814, Jon Berzelius introduced a system of abbreviations for the elements which is still in use today. The symbol was determined by using the first letter of the element plus another letter or letters from the Latin name. Thus, **O** stands for oxygen, **Al** for aluminum, and **Fe** (ferrum) for iron.

IP3412 Chemistry

Name _____   Date _____

EXAMPLES OF ATOMIC SYMBOLS

Element	Atomic number	Alchemist's symbol	Modern symbol	Name from which symbol is derived
Sulfur	16		S	**S**ulfur
Manganese	25		Mn	**M**a**n**ganese
Iron	26		Fe	**Fe**rrum
Cobalt	27		Co	**Co**balt
Nickel	28		Ni	**Ni**ckel
Copper	29		Cu	**Cu**prum
Zinc	30		Zn	**Zn**c
Arsenic	33		As	**As**enic
Silver	47		Ag	**A**r**g**entium
Tin	50		Sn	**S**tan**n**um
Antimony	51		Sb	**S**ti**b**ium
Gold	79		Au	**Au**rum
Mercury	80		Hg	**H**ydrar**g**yrum
Lead	82		Pb	**P**lum**b**um
Bismuth	83		Bi	**Bi**smuth

MP3412 Chemistr

Name _____ Date _____

ATOMIC STRUCTURE

An **atom** is made up of three basic parts: **protons**, **neutrons**, and **electrons**. The protons which have a positive charge (**+**) and the neutrons which have no charge, form the **nucleus** in the center of the atom. The electrons which have a negative charge (**–**), spin in orbits around the nucleus. Each atom has an equal number of protons and electrons. Because the negative charge of the electrons balances the positive charge of the protons, atoms are neutral.

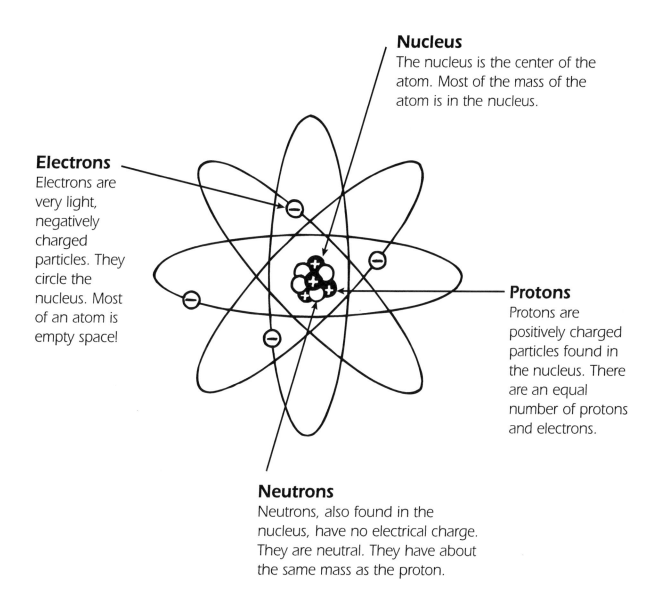

Nucleus
The nucleus is the center of the atom. Most of the mass of the atom is in the nucleus.

Electrons
Electrons are very light, negatively charged particles. They circle the nucleus. Most of an atom is empty space!

Protons
Protons are positively charged particles found in the nucleus. There are an equal number of protons and electrons.

Neutrons
Neutrons, also found in the nucleus, have no electrical charge. They are neutral. They have about the same mass as the proton.

3

Name _____   Date _____

PERIODIC TABLE OF THE ELEMENTS

| Hydrogen
1
H
1.00794(7) | | Key: | | | | | | |

Key:

Element Name
Atomic Number
Element Symbol
2001 atomic weight (mean relative mass)

Lithium 3 **Li** 6.941(2)	Beryllium 4 **Be** 9.012182(3)							
Sodium 11 **Na** 22.989770(2)	Magnesium 12 **Mg** 24.3050(6)							
Potassium 19 **K** 39.0983(1)	Calcium 20 **Ca** 40.078(4)	Scandium 21 **Sc** 44.955910(8)	Titanium 22 **Ti** 47.867(1)	Vanadium 23 **V** 50.9415(1)	Chromium 24 **Cr** 51.9961(6)	Manganese 25 **Mn** 54.938049(9)	Iron 26 **Fe** 55.845(2)	Cobalt 27 **Co** 58.933200(9)
Rubidium 37 **Rb** 85.4678(3)	Strontium 38 **Sr** 87.62(1)	Yttrium 39 **Y** 88.90585(2)	Zirconium 40 **Zr** 91.224(2)	Niobium 41 **Nb** 92.90638(2)	Molybdenum 42 **Mo** 95.94(1)	Technetium 43 **Tc** [98]	Ruthenium 44 **Ru** 101.07(2)	Rhodium 45 **Rh** 102.90550(2)
Cesium 55 **Cs** 132.90545(2)	Barium 56 **Ba** 137.327(7)	57-70 *	Lutetium 71 **Lu** 174.967(1)	Hafnium 72 **Hf** 178.49(2)	Tantalum 73 **Ta** 180.9479(1)	Tungsten 74 **W** 183.84(1)	Rhenium 75 **Re** 186.207(1)	Osmium 76 **Os** 190.23(3)	Iridium 77 **Ir** 192.217(3)
Francium 87 **Fr** [223]	Radium 88 **Ra** [226]	89-102 **	Lawrencium 103 **Lr** [262]	Rutherfordium 104 **Rf** [261]	Dubnium 105 **Db** [262]	Seaborgium 106 **Sg** [266]	Bohrium 107 **Bh** [264]	Hassium 108 **Hs** [269]	Meitnerium 109 **Mt** [268]

	Lanthanum 57 **La** 138.9055(2)	Cerium 58 **Ce** 140.116(1)	Praseodymium 59 **Pr** 140.90765(2)	Neodymium 60 **Nd** 144.24(3)	Promethium 61 **Pm** [145]	Samarium 62 **Sm** 150.36(3)	Europium 63 **Eu** 151.964(1)
*lanthanoids							
**actinoids	Actinium 89 **Ac** [227]	Thorium 90 **Th** 232.0381(1)	Protactinium 91 **Pa** 231.03588(2)	Uranium 92 **U** 238.0289(3)	Neptunium 93 **Np** [237]	Plutonium 94 **Pu** [244]	Americium 95 **Am** [243]

MP3412 Chemistr

Name ______________________________ Date ____________________

PERIODIC TABLE OF THE ELEMENTS

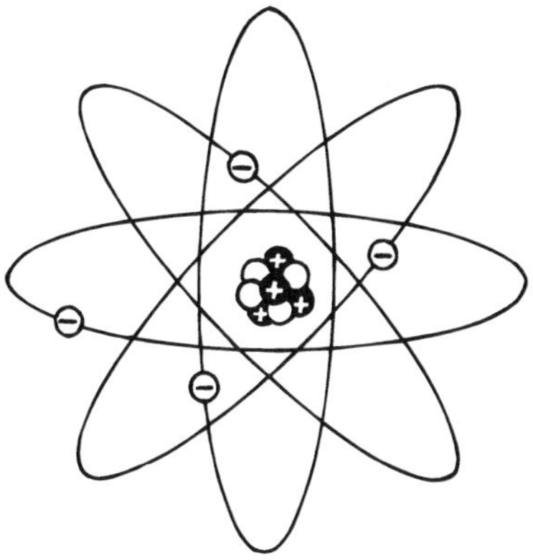

									Helium 2 **He** 4.002602(2)
			Boron 5 **B** 10.811(7)	Carbon 6 **C** 12.0107(8)	Nitrogen 7 **N** 14.00674(7)	Oxygen 8 **O** 15.9994(3)	Fluorine 9 **F** 18.9984032(5)	Neon 10 **Ne** 20.1797(6)
			Aluminum 13 **Al** 26.981538(2)	Silicon 14 **Si** 28.0855(3)	Phosphorus 15 **P** 30.973761(2)	Sulfur 16 **S** 32.065(5)	Chlorine 17 **Cl** 35.453(2)	Argon 18 **Ar** 39.948(1)
Nickel 28 **Ni** 58.6934(2)	Copper 29 **Cu** 63.546(3)	Zinc 30 **Zn** 65.409(4)	Gallium 31 **Ga** 69.723(1)	Germanium 32 **Ge** 72.64(1)	Arsenic 33 **As** 74.92160(2)	Selenium 34 **Se** 78.96(3)	Bromine 35 **Br** 79.904(1)	Krypton 36 **Kr** 83.798(2)
Palladium 46 **Pd** 106.42(1)	Silver 47 **Ag** 107.8682(2)	Cadmium 48 **Cd** 112.411(8)	Indium 49 **In** 114.818(3)	Tin 50 **Sn** 118.710(7)	Antimony 51 **Sb** 121.760(1)	Tellurium 52 **Te** 127.60(3)	Iodine 53 **I** 126.90447(3)	Xenon 54 **Xe** 131.293(6)
Platinum 78 **Pt** 195.078(2)	Gold 79 **Au** 196.96655(2)	Mercury 80 **Hg** 200.59(2)	Thallium 81 **Tl** 204.3833(2)	Lead 82 **Pb** 207.2(1)	Bismuth 83 **Bi** 208.98038(2)	Polonium 84 **Po** [209]	Astatine 85 **At** [210]	Radon 86 **Rn** [222]
Darmstadtium 110 **Ds** [271]	Unununium 111 **Uuu** [272]	Ununbium 112 **Uub** [285]		Ununquadium 114 **Uuq** [289]				

Charts link here.

Gadolinium 64 **Gd** 157.25(3)	Terbium 65 **Tb** 158.92534(2)	Dysprosium 66 **Dy** 162.500(1)	Holmium 67 **Ho** 164.93032(2)	Erbium 68 **Er** 167.259(3)	Thulium 69 **Tm** 168.93421(2)	Ytterbium 70 **Yb** 173.04(3)
Cerium 96 **Cm** [247]	Berkelium 97 **Bk** [247]	Californium 98 **Cf** [251]	Einsteinium 99 **Es** [252]	Fermium 100 **Fm** [257]	Mendelevium 101 **Md** [258]	Nobelium 102 **No** [259]

ATOMIC NUMBERS

The **atomic number** of an element comes from the number of protons that an element has in its nucleus. Each atom of an element has a specific number of protons and electrons. No two elements are alike in this way. So, the atomic number 1 tells us an atom has just one proton in its nucleus. That means it is a hydrogen atom. The atomic number 47 tells us that there are 47 protons in an atom. It is, therefore, a silver atom. Look at the chart on page 2.

Organizing and expanding the ideas of earlier scientists, Dmitri Mendeleev, of Russia, in the 1800s, was able to arrange the elements into a systematic chart or table called the Periodic Table of the Elements. The modern **Periodic Table of the Elements** groups the elements by similar properties. It gives the symbol, the atomic number, and the atomic weight of each element,

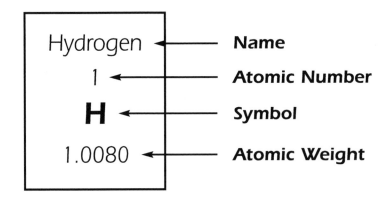

Study the Periodic Table of the Elements on pages 4 and 5. The table shows atomic numbers to 118, Use reference materials to find out about Elements 114, 116, and 118.

_____

_____

_____

_____

_____

_____

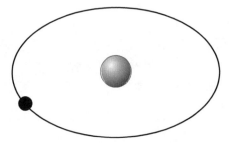

Name _____   Date _____

REVIEW

Complete the sentences. Then find and circle the answers to the questions in the word search.

1. The early Greeks thought that all things were made of _____, _____, _____, and _____.

2. An atom is made of three basic parts: _____, _____, and _____.

3. The _____ is found at the center of the atom.

4. _____ was one of the earliest forms of chemistry.

5. The _____ _____ of an element is determined by the number of protons in the nucleus.

6. The _____ were arranged into a systematic chart by Dmitri Mendeleev.

7. Alchemists tried to make metals, such as lead and iron, into _____.

8. The symbol for an element comes from one or two letters of its English or _____ name.

9. People knew how to produce many _____ changes long before they understood why they happened.

10. The atomic number 1 belongs to the _____ atom.

The words read forwards, backwards, up, down, and diagonally.

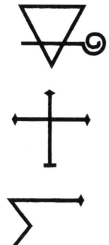

```
P T H J P T T U L A T I N O L N F T Y K B
H W Q I F R G B O U C S U C V R W A M I Z
X T B O L D O J H D E S C H E M I C A L Q
S S T B K I Y T R E T T L N I F N P E D Z
U I V T W M T Y O U J E E I D T E B C X E
V N T R E J S V X N G Y U O L P U B Y R L
A J I H W A P F V C S L S N O R T C E L E
N O C Y B Y I O N F H D D T G E R I N G M
H L D V U M M J W A T E R O N R O E L E E
A B T U O P Y G A I R U Y M M I N D C T N
E T Y G S S V T U I A M G D E F S T V B T
A T O M I C N U M B E R H E W A Q Z T U S
X G I E M F R V Y H Y D R O G E N T E R Z
```

Name _____   Date _____

WHAT IS MATTER?

Chemistry is the study of **matter** and its **changes**. What is matter? Matter is all around you. It is anything that has mass and takes up space. Air is matter, the food you eat is matter, and the chair you are sitting on is matter. You are matter! You see that matter has many different forms. However, in any form, matter is alike in one way. All matter is made up of **molecules**.

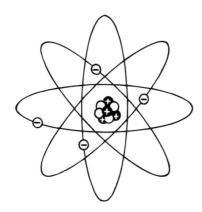

If you took a single grain of salt and crushed it into smaller pieces, it would still be salt. If you could crush it into even smaller pieces, the

smallest piece that would still look, act, and taste like salt would be a salt molecule. The grain of salt you first held is made of billions of salt molecules.

If you could cut a molecule apart, you would discover that it is made of atoms. Table salt is made up of two kinds of atoms, **sodium** and **chlorine**. If someone gave you some **sodium chloride**, you could safely use it on your hamburger and french fries, because sodium chloride is the scientific name for ordinary table salt.

Write the answers in the blanks.

1. The smallest part of a substance that still has the characteristics of that substance is a _____.

2. Chemistry is the study of _____ and its _____.

3. Molecules are made up of _____.

4. You are made of _____.

5. Matter is found _____.

6. Matter has _____ and takes up _____.

7. Table salt is made of two different kinds of _____.

MP3412 Chemistr

Name _____   Date _____

ELEMENTS, COMPOUNDS, MIXTURES

All matter is made up of different kinds of atoms. You might think that there must be millions of different kinds of atoms, but scientists have found just over 100 different kinds. Not all of these occur in nature. Some have been produced only in the laboratory.

Substances that are made up of just one kind of atom are called **elements**. Some familiar elements are oxygen, iron, gold, and hydrogen. Elements may be solids, liquids, or gases. Even though there are more than 100 elements, most matter is made up of just a few of these different atoms. For example, more than 95% of your body is made up of only five elements.

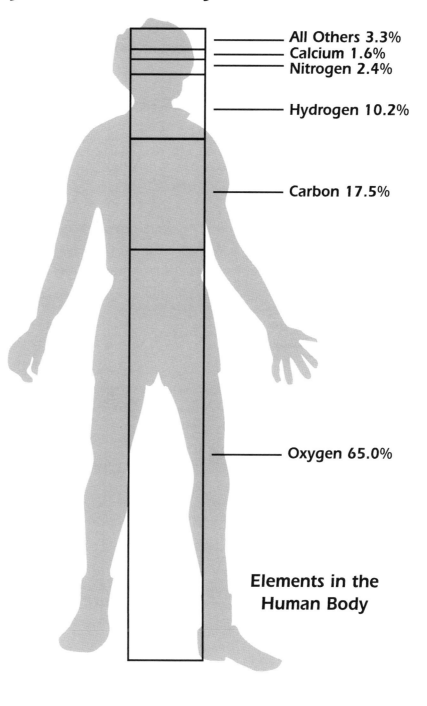

All Others 3.3%
Calcium 1.6%
Nitrogen 2.4%

Hydrogen 10.2%

Carbon 17.5%

Oxygen 65.0%

Elements in the Human Body

ACTIVITY

How can people add calcium to their diets? Use a food chart. List some foods that supply calcium.

_____

_____

_____

Name _____   Date _____

ELEMENTS IN THE EARTH'S CRUST

The Earth's crust, lands below the oceans, is made up of three kinds of rocks. Those rocks are made up mostly of two chemicals which account for more than 75% of the crust's weight. Look at the graph illustrated below. Name the two chemicals. _____ and _____.

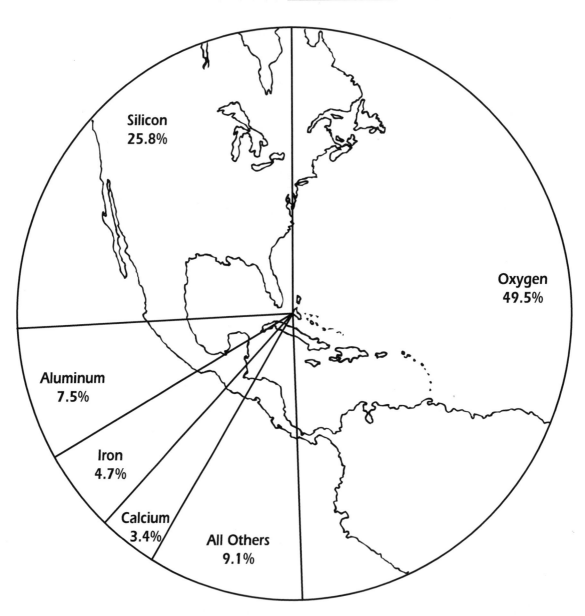

Silicon
25.8%

Oxygen
49.5%

Aluminum
7.5%

Iron
4.7%

Calcium
3.4%

All Others
9.1%

ACTIVITY

Look up silicon. List things which contain this element.

_____

_____

_____

MP3412 Chemistr

Name _____   Date _____

COMPOUNDS AND MIXTURES

If there are so few different kinds of atoms, how do we get so much variety in the matter around us? There are two answers to this question.

1) Atoms are combined in almost endless ways and variations to form different molecules. Although molecules of elements contain atoms of only one kind, molecules of **compounds** are made up of atoms of different kinds. Just as the 26 letters of our alphabet are combined in different ways to form over a million words, the atoms of the elements are combined to form many, many different substances. Some familiar compounds are water and sugar. Water molecules are made of hydrogen and oxygen atoms. Sugar molecules are made of carbon, hydrogen, and oxygen atoms.

2) Some matter is made of different kinds of molecules. Air is a mixture of different kinds of molecules. It contains molecules of oxygen, hydrogen, nitrogen, and other gases. There is no such thing as an "air molecule." Substances made up of different kinds of molecules are called **mixtures**. Wood and gasoline are examples of mixtures.

ACTIVITY

1. Elements are made up of only one kind of _____.

2. The molecules of _____ are made up of different kinds of atoms.

3. Substances made up of two or more different kinds of molecules are called _____.

NOTE

Comparison of Mixtures And Compounds

Compounds	Mixtures
Contain two or more elements	Made up of compounds, elements, or both
Chemically combined	Not chemically combined
Proportion of elements by weight is always the same for a given compound	Amount of elements and compounds will vary
Can be separated by chemical methods	Can be separated by physical methods

MP3412 Chemistry

COMPOSITION OF MATTER: COMPOUNDS

The molecules of compounds are made up of different kinds of atoms. These atoms combine chemically to form a new substance. The following activity will allow you to separate the compound water into its two elements: **hydrogen** and **oxygen**.

MATERIALS

2 carbon rods	2 pieces of 12" insulation wire	paraffin or candle wax	water
2 test tubes	1 bowl or small beaker	6-volt dry cell	

ACTIVITY

1. Remove two carbon rods from two worn-out size D flashlight cells.

2. Strip the insulation from each end of two pieces (each about 12" long) of insulation wire. Wrap one end of the wire around the end of each carbon rod.

3. Melt a small piece of paraffin or candle wax. **Caution!** Always melt paraffin over hot water (as in a double boiler), never directly on a heat element. Dip the end of the carbon rod wrapped with wire into the melted paraffin until it is well coated. The bare wire must be covered with paraffin.

4. A liquid which will conduct electricity is called an **electrolyte**. You will use an electrolyte made with washing soda. Fill two test tubes to the top with the solution. Fill a small bowl or beaker with the rest of the electrolyte solution.

5. Put your thumb or a small square of cardboard over the end of the test tube. Turn the tube upside down and lower it into the bowl. Do not remove your thumb or the cardboard until the open end of the tube is below the fluid level in the bowl.

6. Without letting the solution escape from the tube, slide a carbon rod into each test tube. Connect the wires attached to the carbon rods to the terminals of a 6-volt dry cell.

OBSERVATION

Watch the bubbles collect on the carbon rods. Soon you will see a stream of bubbles rising to the top of the test tube. Gas collecting in the tubes will force the solution out. The gas in one tube is hydrogen and in the other, oxygen.

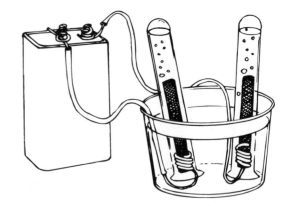

   A. Why do you think there is more of one than the other?

     _____

     _____

MP3412 Chemistry

Name _____   Date _____

COMPOSITION OF MATTER: MIXTURES

Mixtures are substances made up of two or more different kinds of molecules. These molecules keep their own characteristics; they do not chemically combine as the molecules of compounds do. Mixtures are not very easy to separate, but the following activity will give you an idea of some methods that can be used.

MATERIALS

A mixture of equal parts of salt, sand, and iron filings on a piece of paper
water   filter paper   magnet
funnel   2 glass jars or beakers   heat source, if available

ACTIVITY

1. If you had a long time, you could probably separate the iron filings from the salt and sand by picking them out, couldn't you? Can you think of a quicker way to do it?

   _____

2. Try it. Place the separate iron filings on another piece of paper. Explain what you did.

   _____

3. Now you have a mixture of salt and sand. Put the mixture in a glass beaker or jar.

4. Add enough water to the salt/sand mixture to dissolve the salt. Stir.

5. Prepare a piece of filter paper as shown and place it in the funnel. Pour the mixture into the funnel,

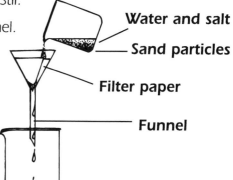

Water and salt

Sand particles

Filter paper

Funnel

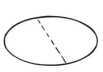

Filter paper   **Fold paper in half.**   **Fold in half again.**   **Make cone from filter paper. Insert into funnel.**

6. Describe what you see. _____

   _____

7. What is left on the filter paper? _____

8. Now where is the salt? _____

9. How can the salt be separated from the liquid? _____

   _____

MP3412 Chemistry

Name _____   Date _____

REVIEW

(1) __ __ __ __ __ __ __ __ __ __

(2) __ __ __ __ __ __ __ __

(3) __ __ __ __ __ __ __ __

(4) __ __ __ __ __ __ __ __ __

(5) __ __ __ __ __ __ __ __

(6) __ __ __ __ __ __ __

(7) __ __ __ __ __ __

(8) __ __ __ __ __ __

(9) __ __ __ __ __ __ __ __ __

Answer the questions below to fill in the blanks above. Write a letter in each blank space. If your answers are correct, you will see a familiar word in the test tube.

1. Atoms combine in many different ways to form _____.

2. Chemistry is the science of matter and its _____.

3. Substances made up of only one kind of atom are called _____.

4. Substances made up of two or more different kinds of atoms are _____.

5. Substances made of two or more different kinds of molecules are _____.

6. Matter has mass and takes up _____.

7. Molecules are made up of _____.

8. All _____ is made up of molecules.

9. Matter is found _____.

10. The word in the test tube is _____.

Name _____   Date _____

ACIDS AND BASES

In our everyday lives, we use many compounds which scientists call **acids**. The orange or grapefruit juice you drink for breakfast contains citric acid (also known as Vitamin C). When milk turns sour, it contains lactic acid. The vinegar you use in your salad dressing contains acetic acid.

Hydrochloric acid and sulfuric acid have many industrial uses. Hydrochloric acid is also produced in small amounts in your stomach where it helps digestion! Strong acids must be handled carefully because they can burn skin, and eyes, and damage clothing.

Scientists call another group of chemicals **bases**. These substances are used often around the home for cleaning because many of them dissolve grease. Ammonia and borax are bases. Milk of Magnesia and baking soda, which may be taken when you have an upset stomach, are bases which act as antacids—they neutralize stomach acids. If you listen to commercials or read the ads for antacid products, this may sound familiar to you! Strong bases must also be handled carefully. Lye is an example of a base that can harm skin, eyes, and clothing.

Substances that are neither acids nor bases are said to be **neutral**. Most cosmetics, shampoos, and products that are used on the skin are neutral,

Scientists use substances called **indicators** to test for acids and bases. These indicators change color in the presence of an acid or base.

1. If you bought a product to clean your kitchen stove, it would probably contain a(n) _____ .

2. Strong _____ and _____ may harm skin and clothes.

3. _____ are found in many common foods.

4. Some _____ are used as antacids.

5. Scientists use _____ to test for acids and bases.

6. Substances that are neither acids nor bases are _____.

Name _____   Date _____

TESTING FOR ACIDS AND BASES

It is often important for chemists to know if a chemical is an acid or a base. Certain substances, called indicators, change color when mixed with acidic or basic substances. You will be using red cabbage juice as an indicator to test a variety of solutions.

MATERIALS

| test liquids | tape | cup of red cabbage juice |
| medicine cups | marker | dropper |

ACTIVITY

1. Fill medicine cups or small jars with about one tablespoon (15 ml) of each liquid provided for you by your teacher. Mark each cup with a small piece of tape. Write the name of the test liquid on the tape.

2. Add about 15 drops (5 ml) of red cabbage indicator solution to each test liquid. Notice the color change. The most acidic liquids will turn red. The most basic solutions will turn yellowish-green. The neutral liquids will not change the color of the indicator.

3. Arrange your cups in order from the most acidic to the most basic. Neutral liquids should be in the middle.

4. List the liquids in order from the most acidic to the most basic.

_____

_____

Scientists have divided the range of acidity into 14 divisions. Each division has been given a number called the **pH** number. A substance with a **pH** of **1** is very acidic. A **pH** of **7** is neutral, and a substance with a **pH** of **14** is very basic.

ACTIVITY

Use the chart on page 17 to make a color line for your test liquids.

MP3412 Chemistry

Name _____   Date _____

COLOR LINE FOR ACIDS AND BASES

Use crayons or colored pencils to match the colors of the liquids you test. Color in a box to represent each liquid after it has been tested. Write the name of the liquid. Use the chart to determine the **pH** of each test liquid. If a liquid is not listed on the chart, try to figure out from its color approximately what its **pH** should be.

SAMPLE pH SCALE

Acidic   **Neutral**   **Basic**

| 1 | 2 | 2.2 | 3 | 3.5 | 4.2 | 5.6 | 7 | 8.2 | 9.0 | 10 | 11 | 13 |

Hydrochloric acid · Lemon juice · Vinegar · Grapefruit juice · Tomato juice · Normal rainfall · Distilled water · Salt · Baking soda · Borax · Milk of Magnesia · Washing soda · Ammonia · Lye

LIQUID TESTED

_____ pH _____ ☐   **Acidic (red)**

_____ pH _____ ☐

_____ pH _____ ☐   **Neutral (purple)**
_____ pH _____ ☐   **Red cabbage juice**

_____ pH _____ ☐

_____ pH _____ ☐   **Basic (yellow-green)**

_____ pH _____ ☐

What do the letters pH stand for? Check your science text or classroom reference materials for the answer.

_____

_____

Name _____ Date _____

TESTING FOR STARCH

You probably know that in order to have a balanced diet you should eat some carbohydrates (sugars, starch, and fiber) each day. Carbohydrates provide vital energy for your body. Scientists test for starch by using another chemical indicator—iodine. Iodine is normally a yellowish-brown liquid, but when a drop of it comes in contact with starch, it turns a deep blue-black. Test a variety of foods to find out which contain starch.

MATERIALS

iodine   dropper
small pieces of several different foods such as apple, cracker, cookie, potato, grapes, cereals, rice, lettuce, celery, etc.

ACTIVITY

1. Place a small drop of iodine on each piece of food. If the iodine turns blue-black, starch is present in the food.
2. Record your findings on the chart.
3. Soak seeds in water for several hours. Cut or split them open and test them for starch.
4. Draw a picture to show which parts of the seed store starch.
5. Why is it important for seeds to contain starch? _____

Caution: Iodine will stain skin and clothing.

Food	Starch	
	Present	Not Present

MP3412 Chemistry

Name _____   Date _____

SOLVENTS, SOLUTIONS, AND CRYSTALS

Water is one of the most common and useful compounds found on Earth. It is certainly very important to people. One of the reasons for water's importance is because it dissolves so many things.

Scientists call a substance that dissolves other things a solvent. The mixture of a dissolved substance **(solute)** and the substance that dissolves it (solvent), is called a **solution**.

MATERIALS

glass jar   hot water   paper clip   thin cardboard
salt   stirring stick   string

ACTIVITY

1. Punch a small hole in the center of the cardboard.

2. Tie a paper clip to one end of the string and thread the other end through the hole in the cardboard. Place the cardboard on top of the jar and adjust the string so that the paper clip is close to, but not touching, the bottom of the jar. Knot the string on top of the cardboard to hold it in place. Set the cardboard and string aside.

3. Fill the glass jar with hot water. Add salt and stir as long as the salt continues to dissolve.

4. When the water will dissolve no more salt, place the cardboard on top of the jar with the paper clip suspended in the salt solution.

5. Place the jar where it will not be disturbed. Watch it for several days. Describe the results.

_____

_____

NOTE

If you remember your work with mixtures, you've probably guessed that as the water evaporates, **crystals** will form. A crystal is a solid in which the atoms are arranged in a particular way. Each type of crystal has its own specific pattern. As your salt crystals grow, you will see that each one has the shape of a **cube**.

MP3412 Chemistry

Name _____   Date _____

IS BLACK ALWAYS BLACK?

Black markers are always black. Right? Well . . . Try the following activity. You may be in for a surprise!

MATERIALS

black felt-tip markers   filter paper   solvents
glass jars   toothpicks   scissors

ACTIVITY

1. Pour about ¾" (2 cm) of each solvent your teacher gives you into a different glass jar. Use a dark crayon to write the name of the solvent on the jar.

2. Cut strips of filter paper narrow enough to fit upright in the jars. They should be a little longer than the jar is deep.

3. About ¾" (2 cm) from the end of the filter paper, make a small dot with the marker you are going to test.

4. Suspend the filter paper in the jar so that the end of the paper is in the solvent. The dot made by the marker should not touch the solvent.

5. Punch a hole in the top end of the filter paper. Put a toothpick through the hole and rest it on the jar rim so that the filter paper is suspended from the toothpick.

6. Be sure you remember which marker you are testing. You may write its name in pencil on the top end of the filter paper if you wish.

7. Watch what happens! Keep records on the chart below.

8. Repeat the process using different markers and solvents.

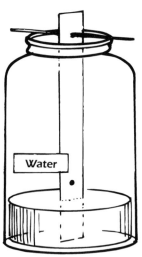

Water

Marker	Solvent	Colors	Distance Traveled

MP3412 Chemistry

Name _____   Date _____

HOW DOES SOAP WORK?

Water is sometimes called the "universal solvent" because it dissolves so many substances. However, not even water dissolves everything.

MATERIALS

a small jar with a lid   water
cooking oil   liquid soap

ACTIVITY

1. Fill the small jar about half full of water.

2. Add about two tablespoons (30 ml) cooking oil to the jar.

   Describe what you see. _____

   _____

3. Put the lid on the jar and shake it. Describe what you see. _____

   _____

4. Let the jar stand for a few minutes. What do you see? _____

   _____

5. Is the water a solvent for oil? _____

6. Now, add a small amount of soap to the oil and water in the jar. Put the lid on the jar and shake it vigorously. Describe what you see.

   _____

7. Can you still see the oil drops? _____

You have made an **emulsion**. Soaps and detergents prevent the oil and water from separating. They are emulsifying agents. Usually dirt clings to you because it contains a small amount of grease. When you use soap to wash yourself, an emulsion is formed of the soap, water, and greasy dirt. The dirt is trapped in the emulsion and is washed away.

MP3412 Chemistry

Name _____   Date _____

REVIEW

ACROSS

1. The mixture of a dissolved substance and the substance that dissolves it is called a _____.
4. Vinegar and lemon juice are examples of _____.
6. _____ is used as an indicator for starch.
8. The _____ scale divides the possible range of acidity into 14 divisions.
10. Foods containing _____ help provide energy for body functions.
11. A substance that dissolves other things is a _____.
12. _____ are often used as household cleaners because many dissolve grease.

DOWN

2. Red cabbage juice may be used as an _____ for acids and bases.
3. Dirt washes away more easily with _____ than with plain water.
5. A _____ is a solid in which the atoms are arranged in a particular way.
7. When soap, oil, and water are shaken up together, an _____ is formed.
9. If substances are neither acidic nor basic, they are _____.

MP3412 Chemistry

SIMPLE CHEMISTRY

OXYGEN AND OXIDATION

As far as people are concerned, oxygen is one of the most important elements in the world. About one fifth of the air we breathe is oxygen. We would not live long without it! Oxygen combines chemically with many substances. Often this is an advantage to us, but sometimes it is not. When oxygen combines with another substance, we call the process **oxidation**.

MATERIALS

narrow glass jar   water
shallow bowl or pan   medium-fine steel wool

ACTIVITY

The following activity will show you an example of oxidation.

1. Dampen the steel wool and push it into the bottom of the jar. Push it in tight enough so that it will stay in place when the jar is turned over, but do not pack it down.

2. Fill the bowl or pan about halfway with water.

3. Place the jar with the steel wool upside down in the bowl or pan. Do not disturb it for several hours.

4. After several hours, look at your experiment. Describe what you see.

   _____

   _____

5. Leave your test set up overnight. Describe what you see the next day.

   _____

   _____

   _____

6. Try to explain why the water has risen in the glass jar.

   _____

   _____

   _____

23

Name _____   Date _____

CANDLE BURNING

Burning is another form of oxidation. When something burns, it combines with the oxygen in the air surrounding it. However, burning, as you know, takes place much more quickly than rusting.

MATERIALS

small glass jar   birthday candle
shallow bowl or pan   oil-base clay
water   long wooden matches

ACTIVITY

1. Using the clay to support it, stand the candle in the center of the bowl or pan.

2. Fill the bowl or pan half full of water.

3. Light the candle. Place the glass jar over the lighted candle.

4. Describe what happened. _____

   _____

5. Repeat the process several times. Are the results always the same? _____

   _____

6. Why do you think the candle went out? _____

   _____

   To find out if the burning process used all the oxygen in the glass, combine this activity with the steel wool activity from page 23.

7. Put damp steel wool in the bottom of the jar.

8. Repeat steps 1–3 with the damp steel wool in the jar you will place over the candle.

9. After the candle goes out, mark the water level in the jar with a marker. Do not disturb the test set up.

10. Check your test set up the next day. Did the water rise more? _____

11. Did the burning candle use up all the oxygen in the jar? _____

12. How do you know? _____

    _____

    _____

MP3412 Chemistry

Name _____   Date _____

AN EXPLOSIVE ACTIVITY!

In an earlier activity, you separated water into two gases—oxygen and hydrogen. In this activity, you will produce a third gas—**carbon dioxide**.

MATERIALS

a narrow-neck bottle with tight fitting cork stopper
vinegar   baking soda
piece of paper towel   petroleum jelly

ACTIVITY

1. Pour about ⅓ cup (75 ml) vinegar into the bottle.

2. Place about 1 teaspoon (5 ml) baking soda on a small square of paper towel. Roll the edges of the paper around the baking soda to make a package small enough to fit through the opening of the bottle. Twist the bottom end of the paper.

3. Rub petroleum jelly on the cork stopper.

4. Drop the baking soda package, twisted end first, into the bottle.

5. Quickly, place the cork firmly in the bottle.

6. Shake the bottle until the baking soda is released from the package. Hold the bottle so the cork is facing away from your self and classmates. Describe what happens.

_____

_____

_____

_____

NOTE

When vinegar and baking soda are mixed, a chemical action takes place and carbon dioxide is released.

Name _____   Date _____

CARBON DIOXIDE

When the element carbon reacts with the element oxygen, a new compound, carbon dioxide, is formed. It is a colorless, odorless gas that does not act like either carbon or oxygen. It has characteristics all its own. Green plants use carbon dioxide during photosynthesis to produce food for the plant. People also use carbon dioxide. Carbon dioxide bubbles produced by yeast or baking powder make cakes and breads rise. Soft drinks also contain carbon dioxide to make them fizz.

Try the following activity to find another way that people use carbon dioxide. You will produce carbon dioxide by mixing baking soda with vinegar.

MATERIALS

glass or jar   baking soda   wide-mouth jar
vinegar   a short candle   long wooden matches

ACTIVITY

1. Place the candle in the wide mouth jar. Light it.

2. Pour about ¼ cup (60 ml) vinegar into the drinking glass.

3. Add about 1 teaspoon (5 ml) baking soda to the vinegar.

4. When some of the fizzing has stopped, carefully "pour" the carbon dioxide that you have produced in the glass over the candle. Be careful not to pour any of the liquid from the glass onto the candle.

5. Describe what happens._____

   _____

6. From this activity, would you say that carbon dioxide is lighter or heavier than air?

   _____

7. Explain.

   _____

   _____

   _____

8. Can you think of a way to use this characteristic of carbon dioxide that you have discovered?

   _____

MP3412 Chemistry

Name _____ Date _____

SOME WAYS WE USE GASES

What gases are being used in each picture? Check your science text or classroom reference materials.

1. _____ 2. _____

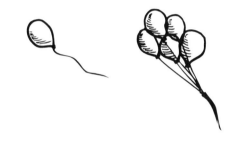

3. _____ 4. _____

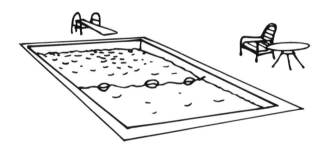

5. _____ 6. _____

7. _____

MP3412 Chemistry

Name _____   Date _____

REVIEW

Complete the sentences.

1. When two or more elements combine to form a new and different substance with characteristics all its own, that new substance is called a _____.

2. Oxygen combines chemically with many substances. This process is known as _____.

3. If vinegar and baking soda are mixed, what gas is produced? _____.

4. The iron in wet steel wool and oxygen combine to form _____.

5. Name two examples of oxidation. _____, _____.

6. _____ is one of the most important elements in the world.

Circle the correct word or words.

7. If a jar is placed over a lighted candle, the flame will go out/stay lighted.

8. Carbon dioxide puts out/spreads a flame.

9. A candle burning in a closed container uses all/some of the oxygen before it goes out.

10. If a jar with wet steel wool is placed upside down in a dish of water, the water will stay the same/rise in the jar.

11. Carbon dioxide bubbles make cakes and breads rise/fall.

12. Carbon dioxide is lighter/heavier than air.

Name four gases and list one way each is used.

_____

_____

_____

_____

MP3412 Chemistry

SIMPLE CHEMISTRY BACKGROUND MATERIAL

Pages 1–6: Students can find out more about recent advances in chemistry, atomic symbols, and the reasons for some of the more recent element names on the Periodic Table through research at the library and the internet.

Pages 8–12: The molecules of compounds are made up of different kinds of atoms. When these atoms combine chemically, they lose their own characteristics. The compound that forms has a new set of characteristics that are uniquely its own. Because compounds are chemically combined, they must be chemically separated. This is often a difficult process. The activity on page 12 is a demonstration of this chemical separation.

An electrolyte can be made with 12 tablespoons of washing soda to 1 quart of water.

Place paraffin or wax in a small can or other disposable container and place in a pan of water to be heated. Never place container of paraffin directly on heat. If students follow the directions carefully, they will see that gases begin to collect very quickly in the tops of the test tubes. They should also notice that twice as much gas collects in one tube as in the other. The larger amount of gas is hydrogen, the smaller, oxygen. Alert students will realize that a water molecule (H_2O) contains twice as much hydrogen as oxygen and will be able to identify which gas is which. Oxygen will be produced by the carbon rod which is connected to the positive terminal of the battery, hydrogen by the rod connected to the negative terminal.

In this classic experiment, the water molecules are broken down, or decomposed, by the electrical current. This method is known as electrolysis.

If students wish to test each of the gases, it may be done with a burning splint or broom straw. This should be done with great care and is probably best done by the teacher as a demonstration. When the tube is full of hydrogen (after about a half hour), remove the carbon rod. Without raising the tube above the solution level, cover the end of the tube with your thumb or a small piece of cardboard. Lift the tube from the bowl and turn it right side up, being careful not to let the hydrogen escape. Bring a

lighted splint or broomstraw close to the test tube. You will hear quite a loud "pop" and perhaps, see a flash as the hydrogen ignites. This should never be done with a larger amount of hydrogen for fear of shattering the glass.

The other test tube should be raised in a like manner. Light a splint or broomstraw and then blow it out so that it is just glowing. Put the glowing splint into the test tube containing the oxygen and the splint will burst into flame.

Page 13: Many substitutions may be made in the materials. Sawdust may be used for sand, tiny snips of steel wool for iron filings, paper towels or coffee filters for filter paper, glass jars instead of Pyrex beakers.

Students should think of using the magnet to separate the iron filings. If Pyrex containers are used, they may be placed over low heat to evaporate the water, leaving the salt in the dish. At that point, all three components of the mixture will have been separated. If Pyrex dishes and heat source are not available, place containers of salt water in a warm, sunny place and wait for natural evaporation to take place.

Pages 15–17: Red cabbage juice indicator solution may be made by cutting up a head of red cabbage, putting it in a pan with enough water to cover it, and boiling for 15 to 20 minutes. This should produce a rich purple liquid. If it is to be kept more than a day or two, it should be refrigerated.

If medicine cups are not available, small jars may be used. The measurements are not critical. If larger jars are used, proportions should be increased.

A selection from the following list of liquids is suggested, but feel free to use your imagination. Add any others you wish to test.

Acids	Bases	Neutrals
white vinegar	ammonia	sugar water
lemon juice	baking soda	water
grapefruit juice	solution	cooking oil
white soda	milk of	salt water
club soda	magnesia	

Other chemicals are often used to test for acids and bases. Red litmus paper turns blue in basic solutions; blue litmus turns red in acids. Universal indicators give a range of colors. Some indicators found in garden shops are specifically designed for use in testing the acidity of soils. Other fruit and vegetable juices, such as beet juice and grape juice, may also be used as indicators.

The pH scale runs from 1 to 14, with 7 considered neutral. Numbers below 7 are acidic and numbers above 7 are basic. The pH scale is logarithmic. Thus, a substance with a pH of 4 is ten times more acidic than a substance with a pH of 5.

Page 18: Crackers, cookies, potatoes, cereals, and rice should all show the presence of starch. Ripe fruits will probably show no evidence of starch, but if you are able to obtain a very green apple, it may show some starch that has not yet been converted to sugar.

Iodine is available from the drug store. If droppers are not available, toothpicks may be used to put a drop of iodine on the test foods.

Seeds split in half should show the presence of starch when tested with iodine. This is stored food that will be used as the young plants develop.

Page 19: Crystals of a particular substance may come in different sizes and colors, but they will always have the same crystalline structure. The atoms will always be arranged in the same pattern.

For edible crystals, use sugar instead of salt. You will have rock candy! Epsom salts and alum, both available at the grocery store, produce crystals, though not edible ones.

Students may wish to experiment with hot and cold water for their solutions. They will discover that hot water will usually dissolve considerably more of a substance than cold water. When hot water has dissolved as much of a substance as it possibly can, the solution is said to be supersaturated.

Students may enjoy making a coal garden. You will need some pieces of coal or charcoal to place in the bottom of a glass or plastic dish. Mix 1/2 cup (120 ml) ammonia and 1 tablespoon (15 ml) laundry bluing together. Then dissolve as much salt in this

mixture as it will hold. Pour solution into dish. Add a few drops of food coloring. Crystals will usually start to grow within an hour or two.

Page 20: The procedure used in this activity is called paper chromatography. It is frequently used in laboratories to separate mixtures into their component parts.

Small-tip markers are easiest to use. Be sure to include a Bic Banana™ black marker if possible – the colors are beautiful! Students will be amazed at the number of colors in "black" markers. Brown markers will also give a wide range of colors.

Suggested solvents are water, rubbing alcohol, white vinegar, and ammonia. Students should be able to name the solvent that works best for each marker (the one that gives the widest color separation).

Coffee filters or white paper towels may be used in place of filter paper.

Other colors may be tested, as well as ball-point pens, water colors, and food coloring.

"Distance Traveled" on chart refers to distance the color traveled.

Page 21: Students may be familiar with the expression, "Oil and water don't mix." Oil, which is less dense than water, floats on the top. Even vigorous shaking will do nothing more than break the oil into little drops; they will not mix. They are immiscible. When emulsifying agents are added, these liquids no longer separate. They become a permanent, or semi-permanent, emulsion. Milk and mayonnaise are familiar emulsions.

Students may wish to put a small amount of oil on their hands and try washing it off with just water. Then have them use soap and they will see how much more successful they are in removing it.

Page 23: Any jar will work, but a narrow one will require less steel wool. Medium (#1) steel wool pads from the hardware store should be cut in pieces. Have students use rubber gloves to avoid getting steel wool "splinters" in their hands. Pull pieces apart so they are fluffy. Water may be tinted with a few drops of food coloring for easier visibility.

When students view their test set ups the next day, they should notice that the steel wool has rusted

30

and the water has risen in the glass jar. Oxygen and iron, in the presence of water, have combined to form iron oxide (rust). As oxygen in the jar is used, the air pressure inside the jar becomes less than the air pressure outside the jar. Normal outside pressure forces water up into the jar. Observant students may notice that the water has filled about one fifth of the jar. The water has taken the place of the oxygen used in the rusting process.

Rusting is an example of relatively slow oxidation; burning is an example of rapid oxidation.

If you wish to test a burning splint or candle in the gas remaining in the jar, you will find that it goes out instantly indicating that the jar no longer contains oxygen.

If students examine the rust, they will find that it is brittle and crumbly. Special paints are used on metal structures to help prevent them from rusting. Students may wish to discuss why this is important. (Rusted metal is brittle and crumbly. It does not have the strength of non-rusted metal.)

Page 24: The explanation of the candle-burning activity is more complex than that of the steel wool activity. If students repeat the candle burning several times, they will notice that the water rise is not always the same. Further experimentation will lead them to discover that the speed with which the jar is lowered over the candle produces the variation in water rise.

In the second part of the activity, students will see that there is rusting and additional water rise the second day, indicating that oxygen was still present in the jar after the burning. Actually, considerable oxygen is needed to support combustion, so the candle goes out before all the oxygen is used.

Thus, part of the water rise in the candle-burning activity is caused by a second factor. As the candle heats the air in the glass jar, it expands and some is forced out of the jar before it reaches the water. When the candle goes out, the air cools and contracts. The water rises to fill the space. This water rise is in addition to the amount which rises to take the place of the oxygen used in the burning process. The speed with which the jar is lowered allows a greater or lesser amount of warm air to escape, causing the varying water rise.

These two factors—cooled air occupying less space and oxygen being used in the burning process—act to decrease the air pressure in the jar. Greater air pressure pressing down on the water outside the jar forces water to rise in the jar.

Page 25: Carbon dioxide is generated when vinegar and baking soda are combined. As the gas builds up in the closed container, the pressure finally blows the cork out.

A soda bottle works well for this activity, but any narrow-neck bottle with a tight fitting cork will do. If you hear a slight hissing sound and the cork does not fly off the bottle when the baking soda and vinegar are mixed, then your cork does not fit tightly enough. The gas is escaping between the cork and the neck of the bottle.

Page 26: Carbon dioxide is a gas that does not burn and generally does not support combustion. It is a heavy gas and can keep oxygen away from a flame, thus causing it to go out. The candle should be shorter than the jar in which it sits. The carbon dioxide is then contained by the jar and results are more dependable.

Some fire extinguishers contain liquid carbon dioxide under high pressure. When the fire extinguisher is opened, the carbon dioxide turns to gas almost immediately. The heavy gas settles down upon the fire and prevents oxygen from reaching it. Students may wish to research various kinds of fire extinguishers.

Page 27: This page illustrates some ways gases are used in everyday products. The page provides students with the opportunity to research the various gases, the products in which they are used, and how they are used, e.g., as a gas with oxygen, helium, carbon dioxide, neon, or in combination with other compounds or elements or in other forms as with nitrogen and fertilizer, hydrogenated vegetable oil, chlorine for water purification. Students may also find other products in which gases are used.

ANSWERS

Page 7

1. earth, fire, water, air
2. protons, neutrons, electrons
3. nucleus
4. Alchemy
5. atomic number
6. elements
7. gold
8. Latin
9. chemical
10. hydrogen

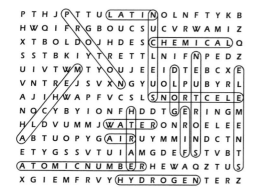

Page 8

1. molecule
2. matter, changes
3. atoms
4. matter (atoms or molecules also acceptable)
5. everywhere
6. mass, space
7. atoms

Page 11

1. atom
2. compounds
3. mixtures

Page 12

A. A water molecule contains twice as much hydrogen as oxygen.

Page 13

1. Use a magnet
2. Put magnet into mixture. Filings cling to magnet and are separated.
6. The liquid goes through the funnel; solid particles are left behind.
7. sand
8. in the liquid in the beaker
9. by evaporation

Page 14

1. molecules
2. changes
3. elements
4. compounds
5. mixtures
6. space
7. atoms
8. matter
9. everywhere
10. chemistry

Page 15

1. base
2. acids, bases
3. Acids
4. bases
5. indicators
6. neutral

Page 17

potential of Hydrogen

Page 21

2. The oil will float on top of the water.
3. Tiny drops of oil are spread throughout the water.
4. The drops of oil begin to rise to the top again.
5. no
6. cloudy or milky liquid, foamy on top
7. no

MP3412 Chemistr